The Essential Guide to AI

What It Is, What It Isn't, and What It Means for Real People

ISBN: **978-1-923567-75-7**

First edition.

Printed and published independently.

For More: **HollySymons.com.au**

Author's Note

Author's Note

I didn't write this book because I'm an AI expert.

I wrote it because I kept watching intelligent, capable people feel overwhelmed, anxious, and quietly inadequate every time AI came up — and I recognised that feeling.

The conversation around AI has become loud, dramatic, and oddly personal. Headlines suggest urgency. Social media rewards certainty. And somewhere in the middle of all that noise, many people have been left feeling like they're already behind in a race they never signed up for.

This book isn't here to turn you into a technologist, a futurist, or an early adopter.

It's here to do something much simpler — and, I think, much more necessary.

It's here to help you understand what AI actually is, what it isn't, and why so much of what's being said about it doesn't quite add up.

You don't need to fear AI.

You don't need to master it.

You don't need to have a position on it at all.

You just need enough clarity to think calmly again.

I've tried to write this book the way I wish the conversation sounded everywhere else: grounded, honest, and human. No hype. No panic. No sales pitch. Just context, perspective, and space to breathe.

If you finish this book feeling steadier than when you started — even if you still have questions — then it's done its job.

And if you decide to put it down partway through and come back later, that's fine too. Nothing here requires urgency.

Understanding was the goal.

Not action.

Not optimisation.

Not keeping up.

Just clarity.

Thank you for being here — and for taking this at a human pace.

The Essential Guide to AI

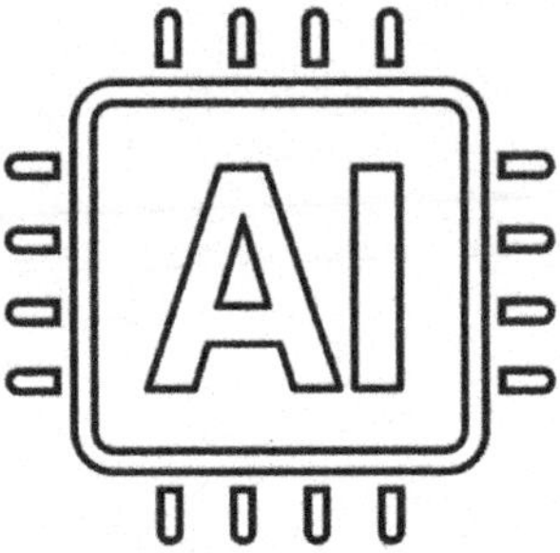

BY HOLLY SYMONS

Table of Contents

Dipping Your Toe in the Water with GPTChat

A gentle, no-pressure introduction for curious but cautious readers

- What GPTChat is (and isn't)
- Tiny step 1: Just say hello
- Tiny step 2: Ask a low-stakes question
- Tiny step 3: Treat it like a draft, not a decision
- Tiny step 4: Try one small, helpful thing
- What you might feel (and why it's normal)
- What you don't need to do
- If you decide it's not for you

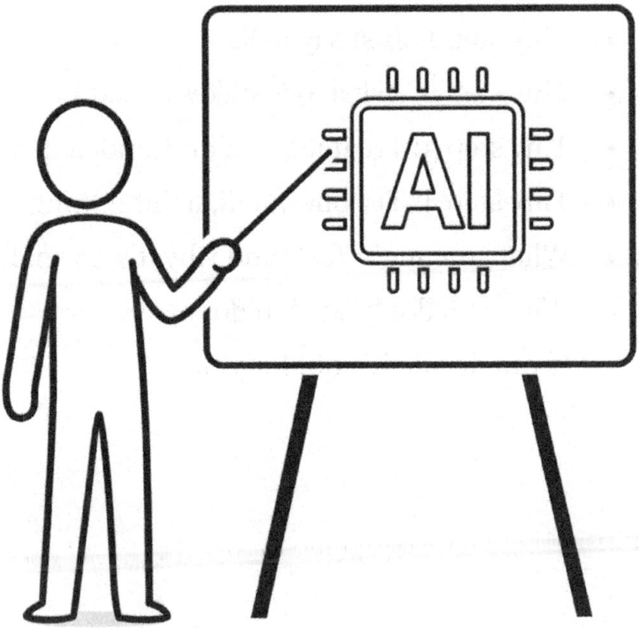
AI

Chapter 1: Your Brain Is Not Broken (AI Just Broke the Internet)

If you've ever read an article about AI and felt:

– vaguely panicked
– strangely behind
– annoyed at people who sound too confident
– suspicious that everyone else understands something you don't

congratulations. You are responding completely normally.

What's happening isn't that you're "bad with technology".

It's that AI arrived in the middle of:
– economic anxiety
– job insecurity
– social media hysteria
– and headlines written by people who get paid per click

Which is a terrible environment for clear thinking.

This book is not here to:

– turn you into a tech expert
– make you love AI
– or tell you everything will be fine

It *is* here to explain what's actually happening — calmly, clearly, and without treating you like an idiot.

Also, just so we're clear:
AI is not a robot.
It does not have thoughts.
It is not planning anything.
And it definitely does not know what it's doing.

We'll get to that.

Nothing is wrong with you.

Chapter 2: Why Everyone Sounds So Confident (and Why You Shouldn't Trust That)

If you've noticed that suddenly *everyone* seems to have strong opinions about AI, you're not imagining it.

People who were recently experts in:

- cryptocurrency
- productivity hacks
- morning routines
- or "manifesting abundance"

have pivoted seamlessly into AI certainty.

This is not because they understand AI deeply.

It's because uncertainty is uncomfortable — and confidence travels better online than honesty.

There's a strange social pressure right now to sound like you've "got it".

To say things like:

- "AI will replace most jobs."
- "AI will change everything."
- "If you don't learn AI, you'll be left behind."

These statements *sound* informative.

They are not.

They're vague enough to be safe, dramatic enough to be shared, and confident enough to silence questions.

And that's the problem.

Real understanding usually sounds slower.

It includes pauses.

It uses phrases like:

- "It depends."
- "In some cases."
- "We don't fully know yet."

Which is terrible for social media, but excellent for thinking.

One of the reasons AI conversations feel so unsettling is that you're often hearing **certainty where uncertainty would be more honest**.

When someone says:

"AI is coming for everyone's job"

what they usually mean is:

"Some tasks are being automated, and we don't yet have good language for what that means emotionally."

That's a very different statement.

Another reason everyone sounds confident is that AI discussions are often framed like a race.

You're either:

- ahead
- behind
- early
- or obsolete

This framing creates urgency — and urgency shuts down curiosity.

If you feel a little resistant when you read AI headlines, that's not ignorance.

That's your brain pushing back against being rushed.

Here's something worth remembering:

Most people talking loudly about AI are not the ones quietly using it well.

The people integrating AI thoughtfully into their work:

- don't talk about it much
- don't use dramatic language
- don't announce themselves as "AI experts"

They treat it like a tool.

Which is... not very exciting to post about.

So, if you've been feeling like everyone else understands something you missed, there's a good chance you're actually responding *more* accurately than they are.

Confusion is an appropriate response to unclear information.

In the next chapter, we'll slow this down properly and look at **the single most important thing to understand about AI** — the one idea that makes everything else easier to think about.

You don't need to be confident yet.

You just need to stay curious.

AI

Chapter 3: AI Is Not a Thing — and That's Where Most of the Confusion Starts

Right now, AI is being talked about as if it's a single, powerful creature.

It has intentions.

It has plans.

It is apparently very busy firing people.

This makes for excellent headlines.

It is also deeply misleading.

When companies announce layoffs and then vaguely mention "AI," it can sound as though a machine has suddenly decided entire roles are no longer needed. But that's almost never how this actually works.

Large technology companies have been restructuring, downsizing, and laying off staff for decades — long before AI entered the public conversation. These cycles often have far more to do with cost control, refocusing priorities, or signalling efficiency to investors than with work "drying up."

This isn't a secret. Employees have joked about it for years.

At IBM, for example, there has long been a piece of dark workplace humour where the company's name is reinterpreted as "I've Been Moved" — a nod to how frequently staff were restructured, reassigned, or laid off during corporate reshuffles. That joke existed well before anyone was blaming AI for job losses.

What *has* changed is the story attached to those decisions.

Today, when layoffs happen, AI is often mentioned as a contributing factor — sometimes alongside reasons that look very similar to past rounds: budget pressure, investor expectations, or organisational change.

That doesn't mean AI plays no role at all.

It does mean the explanation is usually more complex than headlines suggest.

Part of the confusion comes from the way AI itself is described.

AI is talked about as though it *thinks*, *knows*, or *decides*. But AI doesn't do any of those things.

AI predicts.

That's it.

It looks at patterns in data and guesses what comes next. Like an extremely confident autocomplete system that has no idea whether it's right — and no understanding of what it's saying.

This is why AI can sound authoritative while being completely wrong.

And why trusting it blindly is a terrible idea.

It’s also why claims like “AI is making everyone dumber” miss the point.

AI doesn’t make people dumber.

It makes **lazy thinking more visible**.

The same way calculators didn’t destroy maths, but did reveal who never really understood it to begin with.

Then there’s the other kind of noise — the louder, shinier kind.

Teenagers apparently making millions.

Entire careers apparently obsolete.

A constant suggestion that if you're not "mastering AI," you're already falling behind.

This isn't a social shift. It's a marketing one.

If large numbers of people were quietly getting rich using AI, they wouldn't be advertising $5,000 masterclasses about it on social media.

They'd be busy.

The reality is far less dramatic and far more ordinary:

AI is being used to speed up parts of work.

To assist, not replace, decision-making.

And to reduce costs in ways companies were already looking for.

Which is why teachers are still needed.

Lawyers are still needed.

Realtors are still needed.

Marketers are still needed.

And why anyone claiming AI has already replaced entire professions is either oversimplifying — or selling something.

If all of this feels confusing, frustrating, or exhausting, that's not because you're behind.

It's because AI is being talked about in a way that benefits attention, not understanding.

Once you stop thinking of AI as a single "thing" with intentions and start thinking of it as **a collection of tools that predict patterns**, the noise gets quieter.

And from there, you can finally think clearly again.

Confusion fades when you stop mistaking a tool for an actor.

Chapter 4: AI Isn't One Big Thing — It's Lots of Small, Boring Ones

One of the reasons AI feels overwhelming is that it's usually talked about as if it's a single, all-purpose system quietly running everything in the background.

Like there's *one* AI somewhere, getting smarter by the minute.

There isn't.

What we call "AI" is actually a loose collection of very different tools, doing very specific jobs — often badly, sometimes well, and always within limits.

Lumping them together is a bit like calling calculators, spreadsheets, spellcheck, and Google

Maps “the computer” and then worrying that *the computer* is coming for your job.

It’s not wrong exactly — but it’s not helpful.

Most AI tools fall into a few broad categories, and once you see them this way, they stop feeling mysterious.

Some AI systems are built to **recognise patterns**.

They look at large amounts of data and spot trends humans would miss — like fraud detection systems or medical imaging tools. They don’t understand what they’re seeing. They just flag things that look statistically unusual.

Other systems are designed to **predict text, images, or sounds**.

This is where tools like ChatGPT and image generators live. They predict what *usually* comes next based on patterns they’ve seen before. They don’t know if what they’re producing is true, useful, or appropriate — which is why they need humans in the loop.

Then there are systems built for **automation**.

These replace repetitive tasks that follow clear rules — scheduling, sorting, routing, basic customer queries. This kind of automation has existed for decades. It's just been rebranded recently.

Notice what's missing from all of these.

There's no understanding.

No intention.

No awareness of consequences.

That's why it's misleading when people say things like "AI decided" or "AI chose."

AI doesn't choose.

People choose where to use it — and where to trust it.

This is also why AI looks impressive in demonstrations but unreliable in real life. In a demo, conditions are controlled. In the real world, they're messy. And AI is very bad at mess.

If you've ever seen an AI confidently give a wrong answer, misidentify something obvious, or miss basic context, you've already seen its limits in action.

That's not a flaw. It's the nature of the tool.

Understanding this changes the question from:

"What is AI going to do to us?"

to:

"Where are people choosing to use these tools — and why?"

And that's a much calmer question to ask.

Because once you see AI as a set of tools rather than a single force, it becomes easier to:

- question exaggerated claims
- spot misuse
- understand where humans still matter
- and decide how much attention this actually deserves

AI doesn't get to decide the future.

People do.

And they don't always do it particularly well.

Pause here if you need to

You don't have to remember categories.

You don't need definitions.

Just keep the idea that AI is **many tools, not one thing**.

That's enough for now.

AI
?
2010
2015
2020
2020

Chapter 5: If AI Has Been Around for Years, Why Does It Feel Like It Appeared Overnight?

If you're feeling like AI suddenly showed up one day and rearranged the furniture, you're not being dramatic.

It really did *feel* sudden.

One minute, AI was something vaguely mentioned in sci-fi films and corporate strategy documents.

The next minute, it was writing emails, answering questions, making images, and popping up in every news headline.

It's reasonable to ask:

Where was all of this hiding?

The short answer is:

AI didn't suddenly get invented.

It suddenly got **visible**.

For a long time, AI lived quietly in the background of things most people never thought twice about:

- spam filters
- recommendation systems
- fraud detection
- search rankings
- photo tagging

None of these felt personal.

They were useful, mostly invisible, and easy to ignore.

Then something changed.

AI crossed from **back-end systems** into **front-facing tools**.

Instead of quietly sorting things for companies, it started talking directly to people.

And when technology starts *speaking*, everything feels different.

Another reason it felt so abrupt is timing.

AI tools became widely accessible during a period when many people were already:

- tired
- financially anxious
- unsure about work
- overloaded with information

So when a new technology arrived saying, "I can do this faster," it landed on already frayed nerves.

That doesn't mean AI caused the anxiety.

It means it arrived in the middle of it.

There's also a very human reason this felt like an overnight explosion:

we are terrible at noticing slow change.

Progress tends to creep along quietly for years and then cross a threshold where it suddenly feels obvious. When that happens, our brains rewrite history and say, *"This came out of nowhere."*

It didn't.

We just weren't paying attention — and we didn't need to.

Media coverage didn't help.

Once AI tools became visible and interactive, they were perfect headline material:

- dramatic
- clickable
- easy to exaggerate

A calm explanation doesn't spread very far online.

A frightening one does.

So, AI wasn't just introduced – it was introduced **loudly**.

This combination – visibility, timing, and noise – is why so many people feel like they woke up one morning behind the curve.

You weren't.

The curve just moved into your line of sight.

Understanding that helps put things back into proportion. It reminds you that:

- sudden visibility isn't the same as sudden dominance
- new tools don't instantly rewrite reality
- and feeling caught off guard doesn't mean you missed something important

It just means the conversation got noisy.

AI

Chapter 6: Is AI Really Taking People's Jobs?

Let's talk about the question everyone is pretending not to ask.

Is AI taking people's jobs?

The honest answer is: sometimes — but not in the way it's usually described.

What's often missing from this conversation is an important distinction between:
• jobs, and
• tasks

Most jobs are made up of many tasks. Some are repetitive. Some require judgment. Some involve people. Some involve responsibility.

AI is very good at certain tasks.
It is very bad at being responsible for outcomes.

That difference matters more than headlines suggest.

When you hear that a company has "replaced workers with AI," what's usually happened is that:
• some repetitive or predictable tasks have been automated
• fewer people are being hired to do the same volume of work
• or roles have been quietly reshaped without being renamed

That's not nothing — but it's also not the same as entire professions disappearing overnight.

A useful way to think about it is this:

AI rarely replaces a person.
It replaces the least human parts of what a person does.

This is why job loss stories often feel confusing or inconsistent. Two people with the same job title can be affected very differently depending on which parts of their role were most automatable.

It's also why claims like "AI is coming for everyone" are technically meaningless.

Everyone does different work.

Another thing worth saying out loud is that job loss didn't begin with AI.

Automation, outsourcing, restructuring, and cost-cutting have been reshaping work for decades. AI didn't invent those pressures — it's just the latest tool being added to an already unstable system.

That doesn't make job loss easier to experience.
But it does mean AI isn't the sole cause of it.

There's also a quieter change happening alongside the louder one.

In many workplaces, AI isn't removing jobs — it's raising expectations.

People are being asked to:
• produce more
• respond faster
• juggle more tasks
• without more pay or support

This can feel just as threatening as replacement, even when no one is technically being "let go."

So, if you're feeling uneasy, it may not be because you're about to be replaced — it may be because the ground rules of work are shifting.

That's a reasonable thing to notice.

What AI hasn't done is remove the need for:
• accountability
• judgment
• ethical responsibility
• dealing with unexpected situations

When something goes wrong, it's still a human who answers for it.

And that's why AI adoption tends to stall or backtrack in places where:
• mistakes are costly
• nuance matters
• or people are affected in real ways
Which is more common than headlines imply.

The takeaway here isn't "don't worry."
It's also not "panic now."

It’s this:

AI is changing how work is done, not erasing the need for people.

And understanding that difference gives you back a bit of solid ground to stand on.

In the next chapter, we’ll look more closely at which kinds of work are actually most affected — and why. Not so you can fear them, but so you can recognise patterns instead of absorbing noise.

AI changes how work is done — not who matters.

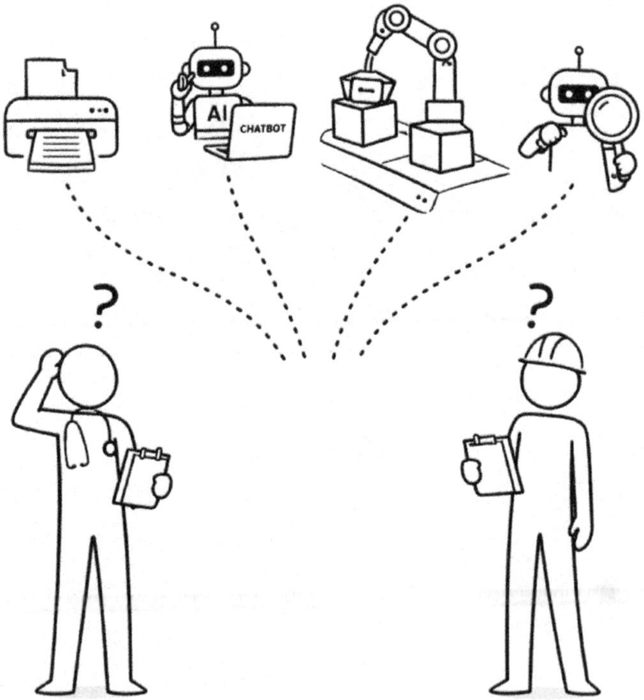
AI
CHATBOT
?
?

Chapter 7: Which Jobs Are Actually Most Affected – and Why

Once you separate *jobs* from *tasks*, a lot of scary claims about AI start to lose their power.

Because it turns out AI isn't especially interested in job titles.

It's interested in **predictability**.

The kinds of work most affected by AI tend to share a few common traits. They usually involve:

- repeating the same steps over and over
- working with structured information
- following clear rules
- producing similar outputs each time

When those conditions exist, automation becomes attractive — not because the work isn't valuable, but because it's **easier to standardise**.

This is why roles built almost entirely around routine processing tend to feel pressure first. Things like:

- basic data entry
- simple reporting
- repetitive administrative tasks
- standard customer responses

Even here, it's rarely the *entire role* that disappears. It's specific slices of it.

On the other end of the spectrum are jobs that rely heavily on:

- judgment
- context
- ethical responsibility
- dealing with people
- adapting to messy, real-world situations

These are much harder to automate — not because they're "better," but because they don't follow neat patterns.

This is why roles involving care, responsibility, negotiation, or accountability tend to change more slowly. When something goes wrong in these areas, there needs to be a human who can explain, adjust, and take responsibility.

AI is very bad at that.

Another group that often feels pressure are roles where **output is valued more than process**.

If the end result matters more than *how* it was produced and quality is judged quickly — AI tools can sometimes be slotted in to speed things up. That doesn't mean the work is gone. It means expectations around speed and volume often increase.

That's an important distinction.

It's also why some people feel *busier* rather than replaced.

One of the more confusing parts of this conversation is watching very different jobs get lumped together as "at risk."

A graphic designer, a lawyer, a teacher, and a real estate agent do wildly different kinds of work — but they all use tools, handle information, and communicate with people. AI may assist parts of each role, but it doesn't replace the whole.

When headlines say "this profession is doomed," what they usually mean is:

"Some tasks within this field are changing."

Which is a much less dramatic sentence — and a much more accurate one.

It's also worth remembering that **new tools don't spread evenly**.

Some industries adopt change quickly.

Others move slowly for good reasons — regulation, ethics, safety, or simple practicality.

This unevenness is why the future doesn't arrive all at once. It arrives in patches.

If you're trying to understand where you personally fit into this picture, a more useful question than "Is my job at risk?" is:

"Which parts of what I do rely on repetition — and which rely on judgment?"

That question leads to insight instead of panic.

In the next chapter, we'll look at the flip side of this — **which kinds of work are least likely to disappear**, and why some roles remain stubbornly human no matter how advanced the tools become.

Chapter 8: Which Jobs Are Least Likely to Disappear (and Why Humans Still Matter)

After reading about which kinds of work are most affected by AI, it's natural to want the opposite list.

The reassurance list.

But instead of naming specific job titles, it's more helpful to understand **why some work resists automation**, even when powerful tools are available.

Because the reason matters more than the label.

Jobs that are least likely to disappear tend to have one thing in common:

Someone has to be accountable when things go wrong.

AI can generate suggestions.

It can speed things up.

It can assist.

But it cannot be held responsible.

Any role where:

- mistakes have real consequences
- judgment must be explained
- decisions affect people directly
- trust matters

will always require a human somewhere in the loop.

This is why work involving care, safety, ethics, or regulation changes slowly. Not because technology isn't capable, but because **society demands accountability**.

It's also why jobs that involve people — not just information — are harder to automate than headlines suggest.

Teaching, healthcare, law, counselling, negotiation, management, sales, and advisory roles all involve reading situations, adapting in real time, and responding to human behaviour. These are areas where context matters more than pattern.

AI struggles there.

Another important factor is **messiness**.

The real world is rarely neat. Information is incomplete. People contradict themselves. Circumstances change. Unexpected things happen.

AI works best in clean environments with clear rules. The messier the situation, the more human judgment becomes essential.

This is why roles that sit at the intersection of:

- people
- systems
- decision-making

tend to endure, even as tools change around them.

It's also why many jobs don't disappear — they **absorb new tools and keep going**.

Think about how many times work has already changed:

- computers
- email
- spreadsheets
- the internet
- mobile phones

Each shift altered how work was done, but didn't erase the need for people to think, decide, and take responsibility.

AI is another shift like that — faster, louder, and more visible, but still constrained by the same limits.

Onc of the quieter truths in this conversation is that **being human is not a weakness in the workplace**.

It's a requirement.

When something matters, people want:

- explanation
- judgment
- reassurance
- responsibility

AI can support those things, but it can't replace them.

So when you hear claims that "AI will replace everyone," it's worth asking:

Who answers when the AI gets it wrong?

If the answer is “a human,” then humans aren’t going anywhere.

In the next chapter, we’ll turn this understanding into something practical – looking at **how people are actually using AI inside existing roles**, without becoming technical or losing their identity.

When responsibility matters, humans stay in the loop.

AI
AI
AI

Chapter 9: How People Are Actually Using AI at Work (Without Making a Big Deal About It)

One of the strangest things about the AI conversation online is how dramatic it sounds compared to how quietly most people are using these tools in real life.

There are very few announcements.

Very few declarations.

And almost no "AI expert" titles involved.

Most people who use AI at work don't talk about it much at all.

They treat it the way people treated spellcheck, templates, or search engines when those first appeared — as something that helps with parts of the job, not the job itself.

In practice, this usually looks like small, unglamorous changes.

Some people use AI to:

- draft rough first versions of emails or reports
- summarise long documents before reading them properly
- brainstorm options they then refine themselves
- check wording, tone, or structure
- reduce time spent on repetitive admin

None of this replaces judgment.

It just reduces friction.

What's noticeable is that people who use AI well rarely hand over full control. They stay involved. They check the output. They decide what stays and what goes.

They don't trust it blindly — and they don't fear it either.

This is especially true in roles where accuracy, reputation, or trust matters. In these contexts, AI is treated more like an assistant than an authority.

Another quiet pattern is that people tend to use AI **inside the role they already have**, rather than reinventing themselves.

They're not "pivoting into AI."

They're making their existing work a little lighter.

That might mean:

- preparing faster before meetings
- organising ideas before presenting them
- clarifying complex information
- handling routine tasks more efficiently

What doesn't show up as often is the dramatic "overnight transformation" story.

Those stories exist — but they're outliers, and they're usually amplified because they're unusual.

Most change is incremental.

It's also worth noticing who benefits most from using AI as a tool.

It's often people who already:

- understand their work
- know what good output looks like
- can spot mistakes
- are comfortable making decisions

In other words, experience matters.

AI doesn't replace expertise.

It amplifies it.

This is one of the reasons the "you must master AI or be left behind" narrative is so misleading. Mastery isn't about learning prompts or tools. It's about knowing *when* something makes sense — and when it doesn't.

If you're curious about AI, that curiosity is already enough to start. You don't need a new identity, a new title, or a dramatic reinvention.

You just need to understand what the tool can do — and what it shouldn't be trusted to do.

In the next chapter, we'll look more closely at **what AI is good at and what it's bad at**, because knowing the limits is often more empowering than knowing the features.

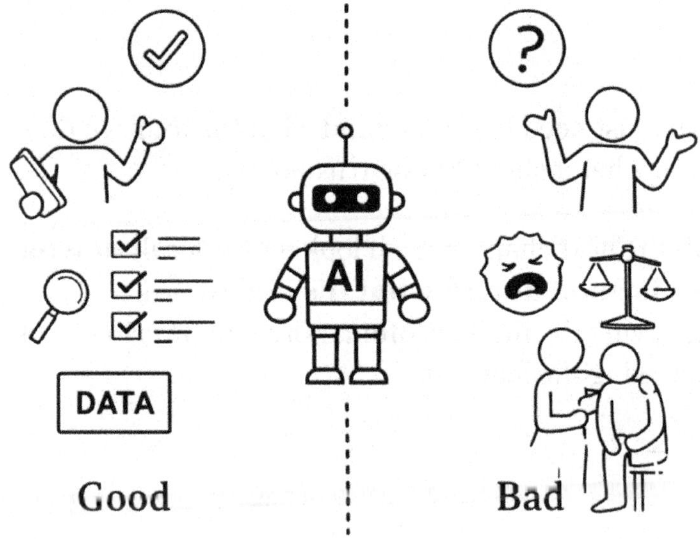
DATA
AI
Good
Bad

Chapter 10: What AI Is Good At — and What It's Bad At (This Matters More Than You Think)

One of the most empowering things you can understand about AI is that it has a very uneven skill set.

It's impressive in some areas.

And surprisingly poor in others.

Knowing the difference makes it much easier to decide when to use it — and when to ignore it.

AI is generally very good at:

- spotting patterns in large amounts of data
- generating drafts or first passes
- summarising information
- producing variations on a theme

- working quickly without getting tired

These strengths are why AI can feel magical at first. It responds instantly, sounds confident, and produces something that looks complete.

But that confidence is deceptive.

Because AI is also very bad at:

- understanding context
- knowing when something *shouldn't* be done
- recognising nuance
- judging consequences
- taking responsibility for mistakes

It doesn't know when it's wrong.

If you've ever had AI give you an answer that sounded convincing but didn't quite make sense, you've already seen this limitation in action. AI isn't lying — it simply doesn't know the difference between "sounds right" and "is right."

This is why AI can be useful for starting work, but risky for finishing it.

It can help you get unstuck.

It can't decide what matters.

Another important weakness is that AI doesn't understand people.

It doesn't read the room.

It doesn't notice discomfort.

It doesn't pick up on unspoken cues.

It doesn't care if something lands badly.

Any task that involves:

- persuasion
- trust
- empathy
- judgment
- or accountability

still requires a human presence.

This is also why AI often struggles in messy, real-world situations. The moment things deviate from the expected pattern, it becomes unreliable.

Humans are good at handling exceptions.

AI is not.

Understanding this changes how you relate to the tool.

Instead of asking:

“Can AI do this?”

A better question becomes:

“Should AI be doing this?”

That question puts *you* back in control.

It also explains why people who use AI effectively tend to be calm about it. They’re not dazzled by its strengths, and they’re not afraid of its weaknesses. They know where it helps — and where it doesn’t belong.

If there’s one idea worth carrying forward from this chapter, it’s this:

AI is powerful where rules are clear.
Humans are essential where meaning, judgment, and responsibility matter.

That’s not a temporary gap.

It’s a structural one.

In the next chapter, we’ll use this understanding to look at how people can work **with** AI without losing their voice, their judgment, or their sense of value.

AI

Chapter 11: How to Use AI Without Losing Your Voice (or Your Judgment)

One of the quieter fears around AI isn't really about jobs.

It's about identity.

People worry — often without quite saying it out loud — that if they start using AI, their work will stop feeling like *theirs*. That everything will sound the same. That they'll lose the part of their work that actually feels human.

That fear makes sense.

Because AI is very good at producing something that looks finished, even when it isn't *right*.

The key to using AI without losing your voice is understanding **where it belongs in the process**.

AI works best at the beginning.

It's good at:

- getting ideas moving
- breaking blank-page paralysis
- generating rough options
- organising thoughts

It's much worse at:

- deciding what matters
- choosing tone
- knowing what to leave out
- understanding what *you* care about

That means the most sensible way to use AI is as a **starting point**, not a final authority.

People who stay comfortable using AI tend to follow a simple pattern:

- they ask for help early
- they take what's useful
- they discard the rest
- and they make the final call themselves

They don't outsource judgment.

Another helpful rule of thumb is this:

If you wouldn't trust a junior assistant to make the final decision, you shouldn't trust AI to do it either.

AI doesn't know your audience.

It doesn't know your values.

It doesn't know the consequences of getting it wrong.

You do.

That's why your role doesn't disappear when you use AI — it becomes clearer.

You're the editor.

The decision-maker.

The person who knows when something sounds right.

This is also why it's okay if AI output feels generic at first.

That's not a failure.

That's the default.

AI produces averages.

Humans produce meaning.

The value comes from what *you* add:

- context
- nuance

- lived experience
- judgment

Using AI doesn't mean becoming less human at work.

It means being more deliberate about what only humans can do.

If you ever find yourself feeling uneasy about relying on AI, that discomfort is often a sign that you're still paying attention — which is a good thing.

Discomfort can be a guide.

In the next chapter, we'll look at how fear around AI gets amplified — and how to spot the scams, hype, and bad advice that thrive on that fear.

NEXT
BIG
THING
AI
FREE!
?
$

Chapter 12: How to Spot AI Scams, Hype, and Bad Advice (Without Becoming Cynical)

Any time there's fear, uncertainty, and rapid change, a second industry appears almost immediately.

It's not technology.

It's reassurance-for-sale.

AI has created perfect conditions for this:

people are worried about work, unsure what to learn, and afraid of falling behind. That makes bold promises very tempting.

The good news is that most bad advice follows a few predictable patterns. Once you see them, they become much easier to ignore.

One common pattern is **false urgency**.

If someone tells you:

- “You must act now”
- “This won’t work in six months”
- “Everyone else is already doing this”

they’re usually trying to bypass your judgment.

Real change doesn’t require panic.

And genuine opportunities don’t evaporate overnight.

Another red flag is **vague success stories**.

Be cautious of claims that focus heavily on:

- income numbers without context
- timelines without detail
- personal wealth without explaining the work involved

Statements like “I made a million dollars using AI” are meaningless without specifics — and specifics are usually where the story falls apart.

Then there’s the **mastery myth**.

Anyone claiming to have “mastered AI” is misunderstanding what AI actually is.

AI tools change constantly.

There is no final level.

No secret prompt.

No permanent advantage.

When someone sells certainty in a rapidly changing space, they’re not selling knowledge — they’re selling comfort.

Another useful question to ask is:

"Who benefits if I believe this?"

If the answer is "the person selling me something," it's worth slowing down.

This doesn't mean everyone talking about AI is dishonest.

It does mean that fear is often monetised before understanding has time to catch up.

There's also a quieter kind of bad advice that sounds reasonable on the surface.

Advice like:

- "You should change careers immediately"
- "Your skills are obsolete"
- "This profession is doomed"

These statements usually ignore how work actually changes — gradually, unevenly, and differently for different people.

They also strip people of agency.

Good advice expands your options.

Bad advice narrows them.

One of the most grounding things you can remember is this:

If advice makes you feel smaller, more anxious, or less capable — it's probably not good advice.

Understanding AI doesn't require buying access, rushing decisions, or reinventing yourself on someone else's timeline.

It requires:

- curiosity
- patience
- and the ability to think clearly under noise

You already have those.

In the next chapter, we'll step back from tools and talk about something more important: **why fear is such a bad guide for career decisions**, and what tends to work better instead.

Good advice leaves you feeling clearer — not smaller

AI
DEAD
END
NEW
PATH

Chapter 13: Why Fear Is a Terrible Career Planner (and What Works Better)

Fear is very good at one thing.

It gets your attention.

It is much worse at helping you make decisions.

When people feel threatened — by technology, money worries, or uncertainty about the future — the brain switches into survival mode. In that state, everything feels urgent, binary, and personal.

You're either:

- ahead or behind
- safe or doomed
- adapting or failing

This framing is emotionally powerful — and strategically useless.

Fear shortens your time horizon.

It pushes you toward drastic moves.

And it makes other people's confidence look like certainty.

That's why career advice given during periods of panic is often extreme:

- "Quit now."
- "Pivot immediately."
- "This field is dead."
- "You're running out of time."

In reality, careers almost never change that way.

They shift gradually.

They overlap.

They absorb new tools and expectations.

Most people don't wake up one morning obsolete. They slowly adjust what they do, how they do it, and where they focus — often without giving it a dramatic name.

Fear doesn't see that process.

It only sees threat.

Another problem with fear-driven planning is that it treats everyone as interchangeable.

But people aren't starting from the same place.

They don't have the same experience.

They don't value the same things.

They don't face the same constraints.

Advice that ignores this isn't just unhelpful — it's destabilising.

A calmer, more reliable way to think about work during change is to focus on **continuity instead of disruption**.

Ask:

- What parts of what I do already work?
- What parts require judgment, context, or trust?
- What parts feel repetitive or draining?

Those questions don't demand instant answers.

They invite observation.

Fear says: *"You must decide now."*

Clarity says: *"Let's understand first."*

It's also worth remembering that no technology shift has ever rewarded the most anxious people.

The people who adapt best are usually:

- curious, not frantic
- willing to experiment quietly
- realistic about limits
- patient with uncertainty

They don't rush to declare themselves "future-proof."

They just keep learning at a human pace.

If you've been feeling stuck, frozen, or overwhelmed by the AI conversation, that doesn't mean you're failing to adapt.

It means your nervous system is doing exactly what it evolved to do when information comes too fast and stakes feel high.

The goal isn't to eliminate fear.

It's to stop letting it run the plan.

In the next chapter, we'll look at **how to think about the future without spiralling** — not by predicting outcomes, but by choosing a steadier way to relate to uncertainty itself.

Chapter 14: How to Think About the Future Without Spiralling

Most anxiety about the future doesn't come from not knowing enough.

It comes from trying to know **too much, too soon**.

When people talk about AI, the future is often presented as a single, fast-moving story that everyone is somehow meant to keep up with. One timeline. One outcome. One correct response.

That's not how the future works.

The future doesn't arrive as a finished plan.

It arrives unevenly, inconsistently, and often inconveniently.

Some changes happen quickly.

Others stall.

Some affect certain industries deeply and barely touch others.

Some sound dramatic and change very little.

Spiralling usually starts when we treat every new development as a personal deadline.

A new tool appears →

a headline exaggerates it →

your brain fills in the worst-case scenario →

and suddenly you feel behind in a race you didn't agree to enter.

A steadier way to relate to the future is to stop asking:

"What will happen?"

and start asking:

"What tends to happen in situations like this?"

Historically, big shifts follow a familiar pattern:

- early excitement
- inflated expectations
- uneven adoption
- disappointment in some areas
- quiet integration in others

Most of the time, life continues in the gaps between headlines.

Another helpful shift is to stop treating the future as something you have to *prepare for perfectly*.

There is no perfect preparation.

There is only **ongoing adjustment**.

People who cope best with uncertainty tend to do a few simple things:

- they notice patterns instead of chasing predictions
- they update their thinking slowly
- they don't confuse noise with direction
- they allow plans to remain unfinished

They don't try to "solve" the future.

They stay responsive.

It's also worth being kind to your own limits.

Humans are not built to constantly imagine worst-case futures while also functioning day to day. Feeling overwhelmed doesn't mean you're bad at change — it means you're human.

You don't need a five-year plan.

You don't need a new identity.

You don't need to know where everything is going.

You just need enough clarity to take the **next reasonable step**, when one presents itself.

And sometimes, the next reasonable step is simply understanding what's happening a little better — which you're already doing.

In the next chapter, we'll bring all of this together and talk about **what this actually means for real people**, without abstraction or pressure.

Chapter 15: So, What Does All of This Actually Mean for Real People?

After all the noise, opinions, headlines, and hot takes, it's reasonable to want a simple answer.

Not a slogan.

Not a prediction.

Just something solid enough to stand on.

So here it is.

For most real people, AI does not mean:

- instant job loss
- sudden irrelevance
- or a need to reinvent yourself overnight

What it means, more often than not, is **gradual change at the edges**.

Some tasks get faster.

Some expectations shift.

Some tools become standard.

And most work continues — just slightly differently than before.

This is not a failure to "keep up."

It's how change actually happens.

If you step back and look at what we've covered so far, a few steady truths emerge.

AI is not a single force moving in one direction.

It's a collection of tools, adopted unevenly, with limits that matter.

AI does not understand, decide, or take responsibility.

People still do.

Job titles are far less important than the kinds of tasks involved.

And most jobs contain a mix that can't be fully automated.

Fear thrives in vague language.

Clarity thrives in specifics.

And perhaps most importantly:

You don't need to win some imaginary race to remain valuable.

Real people don't adapt by chasing every trend.

They adapt by:

- paying attention
- adjusting gradually
- learning when it's useful
- and ignoring what doesn't apply

That doesn't make headlines.

But it works.

If AI becomes part of your working life, it will likely arrive quietly — as a tool that helps in small ways, not as a dramatic turning point. And if it doesn't apply to your work at all, that's not a personal failing. It's simply how uneven adoption works.

Understanding AI isn't about becoming enthusiastic or resistant.

It's about being **oriented**.

Once you understand what AI can do, what it can't do, and how it's being talked about, you regain the ability to think clearly again — and that's the real advantage.

Not speed.

Not mastery.

Clarity.

In the final chapters, we'll reinforce this perspective and leave you with a calm, sensible way to move forward — without pressure, hype, or fear.

?
AI

Chapter 16: You Don't Need to Be an Expert to Be Relevant

Somewhere along the way, the conversation about AI picked up an unhelpful assumption:

That relevance belongs only to experts.

That if you're not constantly learning new tools, new terminology, or new systems, you're falling behind.

That relevance is something you earn by keeping up.

This assumption sounds reasonable — and it's wrong.

Most people who remain relevant over long careers don't do it by becoming experts in every new thing.

They do it by **understanding enough** to make good decisions when it matters.

Relevance doesn't come from mastery.

It comes from judgment.

You don't need to know how AI works internally to know when an output doesn't make sense.

You don't need to understand every tool to recognise when something feels inappropriate, risky, or poorly suited to the situation.

You don't need to stay on the cutting edge to notice what actually improves your work — and what just adds noise.

In fact, people who try to chase expertise in everything often end up less grounded, not more.

They're busy keeping up with tools instead of focusing on outcomes.

They're learning features instead of strengthening judgment.

That's not relevance.

That's exhaustion.

What tends to matter more — especially as work becomes more complex — is the ability to:

- ask good questions
- notice when something doesn't fit
- explain decisions to other humans
- take responsibility when things go wrong

These are not technical skills.

They're human ones.

And they don't expire every time a new tool appears.

This is why people with experience often adapt more smoothly than people with purely technical knowledge. They've seen change before. They know that tools come and go, but judgment accumulates.

If you're someone who:

- thinks carefully
- notices patterns
- questions exaggerated claims
- values clarity over speed

you're already doing the kind of work that remains useful.

You don't need to brand yourself.

You don't need to reinvent your identity.

You don't need to turn curiosity into a performance.

You just need to stay oriented.

AI doesn't raise the bar for being human.

If anything, it makes the human parts of work more visible.

In the final chapter, we'll talk about **a sensible, human way to move forward** — not as a strategy, but as a posture you can return to whenever the noise ramps up again.

Relevance comes from judgment, not expertise.

AI

Chapter 17: A Sensible, Human Way to Move Forward With AI

If there's one thing worth taking with you from this book, it's this:

You don't need a position on AI.

You need a **relationship** with it.

Positions are rigid.

Relationships adapt.

A sensible way to move forward with AI isn't about enthusiasm or resistance. It's about knowing enough to stay oriented — and giving yourself permission to ignore the rest.

That means:

- noticing where AI is genuinely useful
- being cautious where it isn't
- and not mistaking volume for importance

You don't have to adopt every tool.

You don't have to reject them all either.

You get to choose — slowly, thoughtfully, and on your own terms.

A human way forward keeps a few simple ideas in mind.

First:

Technology changes faster than people do — and that's okay.

You're not meant to absorb everything at once. Understanding arrives in layers, and most of it becomes relevant only when it actually touches your life.

Second:

Noise is not direction.

Most of what you'll hear about AI is exaggerated, incomplete, or shaped by someone else's incentives. You don't need to track every development to remain informed. You just need to recognise when something genuinely affects you — and when it doesn't.

Third:

Judgment ages better than skills.

Tools evolve. Interfaces change. Buzzwords expire.

But the ability to think clearly, ask good questions, and notice when something doesn't feel right only becomes more valuable over time.

Fourth:

You're allowed to take this at a human pace.

There is no finish line.

No perfect preparation.

No moment where you suddenly "catch up."

There is just ongoing adjustment — which you're already capable of.

If at any point the conversation around AI starts to feel overwhelming again, you can come back to a few grounding truths:

- AI predicts; it doesn't understand
- tools don't make decisions — people do
- accountability still belongs to humans
- and fear is rarely a good guide

You don't need to be ahead.

You don't need to be early.

You don't need to be exceptional.

You just need to be present, thoughtful, and willing to learn what actually matters when it matters.

That's not falling behind.

That's being human.

You don't need to outrun the future — you just need to stay human in it.

BONUS
SECTION

Bonus Section

Bonus Section: Dipping Your Toe in the Water with GPTChat

A Gentle, No-Pressure Introduction

This section is optional.
You don't need it to understand the book.
You don't need to finish it.
And you don't need to do anything perfectly.

This is just here for people who feel *curious but cautious* — and want to replace fear with familiarity.

Before We Start: What GPTChat Is (and Isn't)

GPTChat is a text-based tool.

That's it.

It's not a robot.

It's not conscious.

It doesn’t know who you are.

It doesn’t remember you once the conversation ends.

It doesn’t “understand” your question — it predicts what text usually comes next based on patterns it has seen before.

That’s why:

- it can sound confident and still be wrong
- it can be useful without being authoritative
- and you don’t need to treat it seriously

You cannot offend it.

You cannot disappoint it.

You cannot break it.

If at any point you don’t like what it says, you can close the window and nothing bad happens.

That’s the whole risk profile.

Nothing bad happens if you stop.

Hello!

Tiny Step 1: Just Say Hello

If you decide to open GPTChat, the screen will show a blank box.

This is the most intimidating part — and it doesn't need to be.

You don't need a clever prompt.

You don't need a goal.

You don't need to know what you're doing.

You can type something as simple as:

"Hello — I'm just testing this."

That's it.

There is no "correct" first message.

There is no score.

This is not a test of intelligence.

You've already succeeded by typing anything at all.

That already counts.

What's a good book to read this weekend?

Tiny Step 2: Ask a Neutral, Low-Stakes Question

Once you see that the tool responds, it helps to ask something that **doesn't matter very much**.

Not a career question.

Not a life decision.

Not something emotionally loaded.

Good low-stakes examples are things like:

- "Can you explain this topic simply?"
- "Can you summarise this article for me?"
- "Can you reword this email to sound more polite?"

The goal here is not to *use* the answer.

The goal is to notice:

- how it responds
- what it gets right
- what it gets wrong

You're observing, not committing.

You are allowed to ignore everything it says.

You’re watching not agreeing.

KEEP
CHANGE
The Alchemist is a great weekend read!
AI

Tiny Step 3: Treat the Response Like a Draft, Not a Decision

This is the most important mindset shift.

GPTChat is best used the way you'd use a rough draft, a brainstorm, or a starting point.

Not an authority.

Not a final answer.

Not something you have to follow.

A helpful question to ask yourself is:

"What, if anything, is useful here?"

You can:

- keep one sentence
- ignore the rest
- reword everything
- or discard it entirely

That's not "doing it wrong."

That *is* the skill.

You decide what survives.

AI

Tiny Step 4: Try One Thing That Makes Life Slightly Easier

If you want to go one step further, try using GPTChat for something **small and practical** – not impressive.

For example:

- organising a messy list of thoughts
- turning a paragraph into bullet points
- clarifying something confusing
- helping you think through options

Nothing about money.

Nothing about optimisation.

Nothing about “scaling”.

Just one small reduction in friction.

If it helps a little, that's a win.

If it doesn't, you've learned something — also a win.

What You Might Feel (This Is Normal)

People often expect a big emotional reaction.

In reality, many people feel:

- mildly curious
- slightly underwhelmed
- a bit uncomfortable
- or unsure what they think

Some feel relief.

Some feel resistance.

Some feel nothing much at all.

All of those responses are normal.

There is no correct emotional outcome.

Learning what doesn't help is still learning.

What You Don't Need to Do

- learn prompts
- become "good at AI"
- keep using it
- talk about it
- or turn this into a habit

Trying something once is enough.

Understanding does not require adoption.

What You Don't Need to Do

It's worth saying this clearly:

You do **not** need to:

- learn prompts
- become "good at AI"
- keep using it
- talk about it
- or turn this into a habit

Trying something once is enough.

Understanding does not require adoption.

AI

If You Decide It's Not for You

That is a valid outcome.

Choosing not to use a tool after understanding it is still a choice — not a failure, not avoidance, not being "left behind."

Plenty of people understand AI perfectly well and decide it doesn't add value to their lives.

That's not resistance.

That's discernment.

One Last Thought

One Last Thought

Fear thrives in unfamiliarity.

Confidence doesn't come from mastery – it comes from knowing *what something is and isn't.*

If this section helped turn GPTChat from a vague, intimidating idea into something ordinary and bounded, it's done its job.

You don't need to go any further than that.

About the Author

Holly Symons writes about complex topics in a way that prioritises clarity, calm, and human understanding.

She is not interested in hype, panic, or selling certainty in uncertain times. Her work focuses on helping people make sense of confusing systems – especially when public conversations become loud, polarised, or anxiety-inducing.

Rather than approaching AI as a technical problem to be mastered, Holly approaches it as a human one: something that affects how people feel about work, competence, relevance, and the future. Her writing aims to restore perspective, reduce unnecessary fear, and help readers think clearly again.

The Essential Guide to AI was written for people who want understanding without pressure – and reassurance without false promises.

Holly lives in Australia and believes that good explanations should make people feel steadier, not smaller

Made in the USA
Monee, IL
05 November 2026

67911932R00085